The Robocentric Transhumanism Advancement
Foundation Series™

Transhumanistic Solar System

WRITTEN BY ALLEN YOUNG
NARRATED BY ALLEN YOUNG
A TRANSHUMANISTIC ASIAN-AMERICAN MAN

Robocentric.com/Books

Robocentric
Press

Copyright

Copyright © 2022 by Robocentric

All rights reserved. No part of this publication may be reproduced or transmitted in any form or by any means, electronic or mechanical, including photocopying, recording, any information storage and retrieval system, or public performance, without permission in writing from the Publisher.

"By Allen Young" on the title page of this book was rendered by the freeware Blade Runner Movie Font v3.01 created by Phil Steinschneider. The font can be freely downloaded at http://www.steinschneider.com/bladerunner/BRFont.htm.

Disclaimer

All information herein is presented for informational purposes only and offered "as is" without contract, warranty, guarantee or assurance of any kind. No responsibility or liability of any kind is assumed by the Publisher for any consequence that arises from the use of the information herein.

About the Author

Allen Young is a transhumanistic Asian-American man who publicly promotes and advances AI, robotics, human immortality biotech, and artificial nuclear-fusion powered mass scale humanity expansion tech.

Allen Young is currently working on raising capital for his high-tech startup, Robocentric, and completing and commercializing his visual AI and robotics software technologies.

More information about Robocentric is at Robocentric.com/About.

You can invest in Robocentric during its early startup years at Robocentric.com/Investors.

A list of all of Allen Young's books on transhumanism with purchase links is available at Robocentric.com/Books.

Table of Contents

Introduction 1
I. Transhumanistic America on Earth and across the Solar System 3
   1. Transhumanistic American Imperialism and Colonialism on Earth and in Outer Space 5
   2. End Game of Transhumanistic Capitalism 7
   3. Transhumanistic American Capitalists 11
   4. Transhumanistic America Empowerment! 14
II. Human Activities in the Transhumanistic Solar System: Beyond AD 2030 16
   5. Social Human Activities in the Transhumanistic Solar System 18
   6. Economic Human Activities in the Transhumanistic Solar System 20
   7. Sexual Human Activities in the Transhumanistic Solar System 21
   8. Political Human Activities in the Transhumanistic Solar System 23
   9. Competitive Human Activities in the Transhumanistic Solar System 25
   10. Collaborative Human Activities in the Transhumanistic Solar System 27
   11. Creative Human Activities in the Transhumanistic Solar System 28
   12. Procreative Human Activities in the Transhumanistic Solar System 29
Closing Words 30

Transhumanistic Solar System

Introduction

Transhumanism is removing human limitations, and improving the human condition, via advancing artificial intelligence, robotics, human immortality biotech, human genetic engineering biotech, neurotech, and artificial nuclear-fusion powered mass scale outer space humanity expansion tech.

My mission is to complete and commercialize, within the next 30 years, by AD 2052, all the artificial intelligence, robotics, human immortality biotech, and artificial nuclear-fusion reactor powered mass scale outer space humanity expansion tech that I envision, according to my plan specified in my book, *The Future*.

God willing, I will achieve all my goals in advancing transhumanism.

God wants me to keep on publicly promoting and advancing transhumanism as long as he wants me to; he won't free me from advancing transhumanism until he is done and through. So I will keep on advancing transhumanism until God tells me to stop.

I cannot help but be obsessed with advancing transhumanism, which is my raison d'etre, my ultimate reason for being.

I strive to create the transhumanistic human reality that I envision: Artificial intelligence, robotics, human immortality biotech, human genetic engineering, neurotech, and artificial nuclear-fusion powered mass scale outer space humanity expansion tech are the technologies that I envision that will revolutionize the human reality and create the next major, transhumanistic epoch of humankind after the hunter-and-gatherer, agrarian, industrial, and digital epochs of humankind.

In this book, I describe the transhumanistic Solar System that I strive to create, that will have immortal human beings everywhere in the Solar System, with mass-scale outer space human transportation, life-support, and habitat systems enabled by artificial intelligence, robotics, and artificial nuclear-fusion powered mass scale outer space humanity expansion technologies.

I. Transhumanistic America on Earth and across the Solar System

I aim to make America the transhumanistic superpower.

As I specify in my book, *The Future*, I want America to have biotechnologically immortal human soldiers, and billions of artificial nuclear-fusion powered interplanetary military spaceships in the Solar System.

In my view, in order for humanity to massively expand into outer space across the entire Solar System, a nation must provide the outer-space infrastructure, safety, and security: I want America to provide the necessary technological, legal, military, and policing infrastructures for enabling the mass-scale outer space humanity expansion in the Solar System—and I want Robocentric, my transhumanistic American high-tech corporation, to develop, manufacture, and supply all the needed technology products for enabling the transhumanistic Solar System with living humans across the entire Solar System.

Naturally, with the enormous, revolutionary, transhumanistic technologies will come the enormous transhumanistic national power: In my view, one consequence of advancing transhumanism will be America becoming the transhumanistic superpower with its national power reaching and influencing not just the entire Earth but the entire Solar System.

Humanity can expand into outer space in massive numbers, only with America providing the necessary framework for enabling the mass-scale outer space humanity expansion across the entire Solar System.

1. Transhumanistic American Imperialism and Colonialism on Earth and in Outer Space

In my view, making AI, robotics, human immortality biotech, and nuclear-fusion powered outer space tech available in every country is not a good idea, at least not initially.

Some nations are very unstable and very corrupt, and introducing transhumanistic technologies to such nations simply would not make sense.

When a nation becomes technologically far more advanced than other nations, it becomes a dominant, imperialistic nation.

My bet is that as transhumanism advances in America, America will become a transhumanistic empire on Earth and in outer space.

My bet is that when the human genetic engineering biotech and neurotech become cheap and abundant in America, the U.S. military will inevitably invade and conquer unstable nations, and genetically engineer the conquered entire human population and install brain chips in every human individual of the conquered nation to make them well-behaving, productive, and biotechnologically advanced human beings.

My bet is that America will play the role of the transhumanistic global and outer-space empire that advances humanity by biotechnologically and neurotechnologically advancing its conquered subject peoples.

Furthermore, there will be a billion or more autonomous U.S. military spaceships across the Solar System that are controlled by the U.S. military human commanders, to enforce safety and security in the Solar System when humans massively expand into outer space using artificial nuclear-fusion powered mass-scale outer space humanity expansion transportation, mining, manufacturing, construction, communication, and human-habitat technologies.

In my transhumanistic vision, that I strive to realize through building my transhumanistic American high-tech corporation, Robocentric, America becoming the transhumanistic superpower is inevitable.

God willing, I will transform America into the transhumanistic superpower that I envision—by developing and commercializing my artificial intelligence, robotics, human immortality biotech, human genetic engineering biotech, neurotech, and artificial nuclear-fusion powered mass scale outer space humanity expansion tech, according to my plan in my book, *The Future*!

2. End Game of Transhumanistic Capitalism

Through observations and experiences, I've concluded that most people are not superintelligent.

Most people lack the intellectual capability to see the truths and realize what really is.

Most likely, until the human intelligence genetic engineering biotech comes about, that can genetically modify human beings to be superintelligent, the vast swath of humanity will remain constrained by average or sub-par human intelligence.

What is the end game of capitalism?

In particular, what is the end game of transhumanistic capitalism, not just on Earth, but across the entire Solar System, and eventually across the different star systems and galaxies, and the entire Universe?

Most people do not know what the end game of capitalism is although it is thrown at their faces everyday, 24/7 if they live in a capitalistic nation and society.

As I think about expanding the national economies of America and other First World nations—not just on Earth but especially in outer space—through advancing AI, robotics, human immortality biotech, and artificial nuclear-fusion reactor powered mass scale outer space humanity expansion tech, I focus more and more on the end game of capitalism.

The end game of capitalism is making people produce more and consume more goods and services.

At its core, capitalism is nothing more than making people produce and consume more.

How do you make people to produce and consume more?

In order to make people to produce and consume more, you've got to show people what they want, such as pleasure, comfort, sense of security, convenience, higher socioeconomic status, benefit, and even privilege.

Capitalism is a very psychological game, because in order to make people to produce and consume more, you must psychologically manipulate people; you've got to convince people that they are better off when they embrace materialism.

Capitalism doesn't work everywhere because some people don't want to embrace materialism.

When you feel like buying something instead of living a simpler life, that's capitalism at work; when you get a desire to buy something, capitalism is seducing you to spend and make itself work.

Sometimes people get tired of working and earning hard, and spending hard; that's when a recession hits in the capitalistic nations and societies.

So, in order to make the end game of capitalism work, which is making people produce and consume more, capitalists must create incentives for working more, and spending more, in the form of goods and services that have the appearances of benefit.

I am a hardcore capitalist. I'm a hardcore capitalist transhumanistic Asian-American man, because I want better and more goods and services for myself and others.

I advance transhumanism; I advance AI, robotics, human immortality biotech, and artificial nuclear-fusion reactor powered mass scale outer space humanity expansion tech, because I want the brand-new, transhumanistic high tech goods and services that never existed before for myself and others.

The end game of transhumanistic capitalism is making the humans in the First World to live much longer, indefinitely if they want to, through using the human immortality biotech, and making the First World peoples to produce and consume far more goods and services—not just on Earth, but especially across the entire Solar System in mass-scale outer space human habitats—through using artificial intelligence, robotics, and artificial nuclear-fusion powered mass scale outer space humanity expansion tech.

As specified in my book, *The Future*, one of my key goals is to sextuple the American GDP to US$132 trillion by transhumanizing the Earth and outer-space national economies of America and the rest of the First World. God willing, I will achieve that goal.

3. Transhumanistic Capitalists   American

A capitalist must be a rabid wolf, disgusting pig, and lascivious dog in chasing money.

A capitalist's love, passion, and lust must be nothing but money.

A capitalist must know nothing but making more and more money as the meaning of the human existence.

To a capitalist, nothing but money must be his or her God.

To a capitalist, nothing but making more and more money must be beautiful and desirable.

To a capitalist, making more and more money must be the be all and end all.

A capitalist must be a rabid wolf, disgusting pig, and lascivious dog in chasing money.

The morality of a capitalist must be all about money.

If it makes money, then it must be right to a capitalist; if it doesn't make money, then it must be wrong to a capitalist.

The people who help making money must be the right people to a capitalist; the people who don't help making money must be the wrong people to a capitalist.

I am Allen Young; I'm a transhumanistic Asian-American man.

I'm a transhumanistic capitalist. I pursue making money via advancing transhumanism, which is removing human limitations and improving the human condition via advancing AI, robotics, human immortality biotech, and artificial nuclear-fusion powered mass-scale outer space humanity expansion tech, according to my plan in my book, *The Future*.

In order to be a transhumanistic capitalist, one must be obsessed with making money using artificial intelligence, robotics, human immortality biotech, human genetic engineering biotech, neurotech, and artificial nuclear-fusion powered mass scale outer space humanity expansion tech.

A transhumanistic capitalist must ferociously exploit the biotechnologically long-living humans in making money, and harness the immortal humans in increasing economic production and consumption, not just on Earth, but especially in outer space, in the Solar System initially, and eventually across multiple star systems and galaxies, and the entire Universe.

A transhumanistic capitalist must passionately exploit the astronomical amounts of resources in outer space for growing and expanding humanity's economic productions and consumptions on Earth and in outer space.

A transhumanistic capitalist must relentlessly hardness artificial intelligence, robotics, human genetic engineering biotech, and neurotech to enable human workers to be far more creative and productive than ever before, and produce and consume astronomical amounts and types of brand-new goods and services on Earth and in outer space.

To be a transhumanistic capitalist is to exploit the biotechnological human immortality, AI, robotics, and artificial nuclear-fusion powered mass scale outer space humanity expansion tech to build wealth at cosmic proportions across the entire Solar System, across trillions of a trillion star systems and galaxies, and across the entire Universe!

4. Transhumanistic Empowerment! America

In my transhumanistic vision that I pursue, the future of America is to become far more powerful and prosperous through transhumanizing itself.

What is transhumanizing America?

Transhumanizing America is biotechnologically making the American people immortal through the human immortality biotech.

Transhumanizing America is making America's presence pervasive across the entire Solar System and far beyond through the artificial nuclear-fusion powered mass scale outer space humanity expansion technologies.

Transhumanizing America is sextupling the American GDP on Earth and in outer space to US$132 trillion via advancing the science, technology, and capitalism of AI, robotics, biotech, neurotech, nanotech, and outer-space tech.

In my transhumanistic vision that I pursue to realize, America will dominate outer space via artificial nuclear-fusion powered mass scale outer space humanity expansion tech.

In my vision that I pursue to realize, the Americans will have an option to become immortal humans via the human immortality biotech.

My vision is to transhumanize America and the rest of the First World.

Transhumanism is removing human limitations, and improving the human condition, through advancing AI, robotics, human immortality biotech, and artificial nuclear-fusion powered mass scale outer space humanity expansion tech, according to my plan in my book, *The Future*.

I toil to transhumanize America, through building my transhumanistic American high-tech corporation, Robocentric.

God willing, I will empower America through advancing transhumanism!

II. Human Activities in the Transhumanistic Solar System: Beyond AD 2030

So far, ever since the dawn of humanity, human behaviors have been dictated by the human biological clock, the humans being confined to Earth, and the humans having limited manual labor and intelligence.

How will humans behave differently, when the human immortality biotech, the artificial nuclear-fusion powered mass scale outer space humanity expansion tech, robotics, and AI will completely and thoroughly remove the human limitations in lifespan, being confined to Earth, and having limited manual labor and intelligence?

How will humans behave in the transhumanistic epoch of humankind?

In my transhumanistic vision that I pursue to realize, I want humans to think and behave with the knowing that they no longer have a limited lifespan, limited resources, limited mental and manual labor, and limited presence only on Earth.

I strive to enable humans to think and behave in transhumanistic ways, as in the following, in which humans are no longer bound and dictated by the aforementioned human limitations that shackled and chained every human being so far ever since the dawn of humankind.

5. Social Human Activities in the Transhumanistic Solar System

Humans are social creatures.

Excluding the introverted human beings, humans are genetically designed to want the company of other human beings, and feel scared when they are alone.

The human genetic design of seeking other human beings gives the tremendous advantage of humans acting and working together to create collective human benefit.

In my vision of the transhumanistic future of humankind, biotechnologically immortal human beings will socialize even more, and do far more humanly things together, not jut on Earth, but more importantly in outer space.

In the transhumanistic future of humankind, biotechnologically immortal humans will travel together to different planets in big interplanetary, interstellar, and intergalactic spaceships that can transport thousands of people per spaceship in outer space.

In the transhumanistic future of humankind, humans will share and enjoy so much of life they have to live, with other human beings.

Limited human lifespan, as well as old and infirm people, will be things of the past: Eternally youthful, beautiful, and healthy human beings will fraternize with each other, on Earth and in outer space, in gigantic spaceships, spaceship fleets, terraformed planets, other life-supporting planets, and mass-scale outer space human habitats.

In the transhumanistic Solar System that I strive to build—through advancing AI, robotics, human immortality biotech, and artificial nuclear-fusion powered mass scale outer space humanity expansion tech—to be human will be being godlike, and humans will savor their godlike existences with other fellow immortal human beings with eternal human youth, beauty, and health, and with unlimited access to the unlimited resources in outer space!

6. Economic Human Activities in the Transhumanistic Solar System

To be human is to produce and consume.

To be transhumanistic human is to produce and consume across the entire Solar System and beyond.

To be human is to go to, and live in, different places.

To be transhumanistic human is to go to, and live in, different places in the Solar System and beyond.

I strive to create the transhumanistic Solar System with tens of a billion artificial nuclear-fusion powered interplanetary spaceships in operation at any given time.

As specified in my book, *The Future*, I strive to build the transhumanistic Solar System economy with star and planet mining outer-space technologies, mass-scale outer space factory and human-habitat construction technologies, and interplanetary communications systems.

To be human in the transhumanistic Solar System is to produce and consume using the resources of the entire Solar System, and to go and live anywhere in the Solar System where humans can go and live!

7. Sexual Human Activities in the Transhumanistic Solar System

What do you think will happen in the human sexual psychology, when humans get the option to live as immortal human beings via the human immortality biotech, and to go and live anywhere in the Solar System via the artificial nuclear-fusion powered mass scale outer space humanity expansion tech?

In the transhumanistic Solar System I strive to create, I want humans to indefinitely have youthful, beautiful, and healthy human bodies, and live sexually liberated lives.

I want humans to have and enjoy sex in interplanetary spaceships and mass-scale outer space human habitats in the Solar System and beyond.

I want humans to enjoy having sex in outer-space apartments, condominiums, townhouses, and single houses.

I want having sex in interplanetary spaceships while traveling to another planet in the Solar System to be a norm.

Because humans will indefinitely live with youthful, beautiful, and healthy human bodies, rampant human sexual promiscuity in the transhumanistic Solar System and beyond will be inevitable.

The humans who choose to live for centuries, millennia, and beyond, across multiple planets, star systems, and galaxies, will inevitably have numerous sexual partners.

Human beings having several thousand, tens of a thousand, and more sexual partners will be a norm in the transhumanistic Solar System and beyond.

Advancing transhumanism will undoubtedly create far more human sexual activities and variety.

8. Political Human Activities in the Transhumanistic Solar System

Politics is a show with practical purposes.

Politics in the transhumanistic Solar System will be an interplanetary affair.

The different peoples inhabiting the different planets with human life-support systems will constantly fight for their benefit and their gain.

With each human-inhabiting planet possessing the outer-space weaponry to destroy entire planets, the interplanetary human contention and animosity will be uneasy most of the time.

In the transhumanistic Solar System, different peoples living in different locales within the Solar System will contend for more benefits for themselves.

Just like how human politics always has been, the human politics in the transhumanistic Solar System will be ugly, fierce, and scary at times.

Talks of interplanetary wars will happen.

Actual interplanetary wars may happen.

Threats of annihilating an entire planet and its human immortality biotech facilities so that no human being on a planet can survive, will be made.

In the transhumanistic Solar System, politics will be dirty, disgusting, and disturbing as always, but it will happen at an astronomical scale.

Brace for the transhumanistic Solar System politics: It will get astronomically ugly.

9. Competitive Human Activities in the Transhumanistic Solar System

Humans advance through competing with each other.

The human desire to outdo, influence, and dominate others always has been the engine of human progress and advancement.

Humans compete socially, economically, sexually, politically, militarily, creatively, and procreatively.

In the transhumanistic Solar System, humans will strive and struggle to demonstrate how much better they are than others.

To be better than others, or at least appear better than others, the humans in the transhumanistic Solar System will strive and struggle to achieve more and have more.

The transhumanistic Solar System humans will exploit and harness the resources from the multiple planets and the asteroid belt in the Solar System to gain an edge and establish their supremacy.

Outer space military conflicts will happen in the transhumanistic Solar System, with billions of artificial nuclear-fusion powered interplanetary military spaceships.

As I mention in my book, *The Future*, the transhumanistic superpower(s) will brandish their planet annihilation outer-space weaponry to display their power, dominance, and supremacy.

Human competition will become even more fierce in the transhumanistic Solar System, because far more energy and resources will become available through the transhumanistic technologies.

The transhumanistic Solar System will be one scary place, not for weak humans, because the human competition, literally, will happen at an astronomical scale!

10. Collaborative Human Activities in the Transhumanistic Solar System

Humans advance not just by competing against each other, but also by collaborating with each other.

Interplanetary collaboration and cooperation will happen in the transhumanistic Solar System.

Interplanetary trade will be astronomically abundant.

In the transhumanistic Solar System, everyday hundreds of a million or more interplanetary cargo spaceships will carry artificial nuclear-fusion fuel, raw materials, and manufactured goods between planets.

Humans will visit other humans living on other planets with human life-support systems, for travel, leisure, and sightseeing.

Humans on different planets will do business together.

Different sexes of humans living on different planets will meet, fall in love, and have children together.

In the transhumanistic Solar System, the human collaboration will no longer be confined to one planet: It will be across the entire Solar System!

11. Creative Human Activities in the Transhumanistic Solar System

Humans are creative creatures by nature.

What will humans create when they become biotechnologically immortal, and have access to the resources of all the planets, moons, and asteroids in the Solar System?

In the transhumanistic Solar System, humans will create more foods, more extracted raw materials, more manufactured goods, more buildings on Earth and in outer space, more information, more sciences, more technologies, more artworks, more entertainment, and more services.

Human creativity will exponentially explode in the transhumanistic Solar System, because there will be so much more human lifespan, resources, and space than the traditional human existence confined to Earth.

Human creativity will reach an astronomical scale in the transhumanistic Solar System, because an entire star system will fuel human creativity.

Human creativity will become even more breathtaking in the transhumanistic Solar System, because humans will have an access to the resources of the entire Solar System and the biotechnologically unlimited human lifespan!

12. Procreative Human Activities in the Transhumanistic Solar System

To procreate is to be human.

In the transhumanistic Solar System, humans will live hundreds, thousands, or more of years, with eternally young, beautiful, healthy, and sexually desirable human bodies, and produce and raise children on different planets.

Due to the human immortality biotech, humanity's population count will exponentially increase, and humans will quickly populate other planets beyond Earth, initially in the Solar System, and eventually in other star systems and galaxies.

The current human population of eight billion on Earth in AD 2022 will pale in comparison to the transhumanistic human population of trillions of a trillion human beings across multiple planets, stars, and galaxies.

Humanity's procreative agenda encoded in the human genes has only begun: Expect the human drama with hundreds of a billion people across multiple planets in the Solar System in the not-so-distant future!

Closing Words

I advance transhumanism to create a human reality that will be far beyond the short vapid human existence that humans lived so far.

I think I cannot be the only one who thinks there must be something far more than this vapid human existence.

I feel that the human existence so far has been empty and meaningless, because I think there must be much more than this to the human life.

There has to be artificial intelligence, robotics, human immortality biotech, and artificial nuclear-fusion powered mass scale outer space humanity expansion tech.

I advance transhumanism, because I want more in the human reality.

God willing, I will build the fantastic human reality of hundreds of a billion biotechnologically immortal human beings being everywhere in the Solar System, poised to expand into other star systems and galaxies across the entire Universe!

(The end)

(Allen Young's other books are available at Robocentric.com/Books.)

www.ingramcontent.com/pod-product-compliance
Lightning Source LLC
Chambersburg PA
CBHW050321220526
45465CB00005B/2079